Contents

Okra	6
Yam	8
Carrot	10
Brussels Sprout	12
Aubergine	14
Artichoke	16
Red Pepper	18
Leek	20
Sweetcorn	22
Potato	24
Pea	26
Index	28

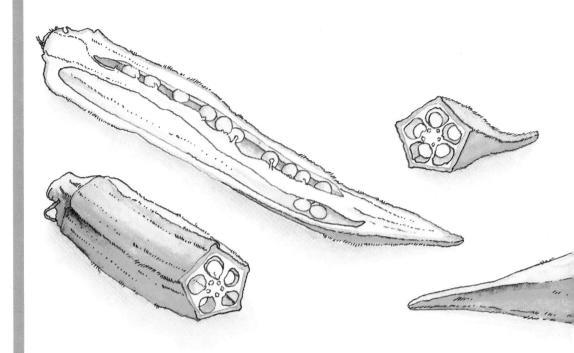

Okra

'Lady's Fingers' is the popular name for okra. This is because the part we eat is long, thin and pointed.

Okra are five-sided green pods, about five to twelve centimetres long. Inside the pods are lots of little brown seeds. The pods grow on green bushy trees with yellow flowers which have a crimson centre. The trees grow

Mixed Vegetables:
A first look at vegetables

Julia Eccleshare
Illustrated by Martin Ursell

Hamish Hamilton · London

> CITY OF BIRMINGHAM
> POLYTECHNIC LIBRARY
>
> BOOK No. 602439
>
> SUBJECT No. 633
>
> Ecc

First published in Great Britain 1986 by
Hamish Hamilton Children's Books
27 Wrights Lane, London W8 5TZ
Copyright © 1986 by Julia Eccleshare (text)
Copyright © 1986 by Martin Ursell (illustrations)
All rights reserved

Designed by Miriam Yarrien

British Library Cataloguing in Publication Data

Eccleshare, Julia
 Mixed vegetables.
 1. Vegetables – Juvenile literature
 I. Title
 641.3'5   TX401

ISBN 0–241–11973–1

Printed in Great Britain by
Cambus Litho, East Kilbride

in hot, wet climates found in parts of India, Africa, the Caribbean and Turkey.

Okra becomes sticky when cooked, and makes a tasty side-dish to eat with a curry.

Yam

Yams come from the West Indies, Central America, and other hot countries.

Above ground, the plant grows green leaves which help it to climb up stakes.

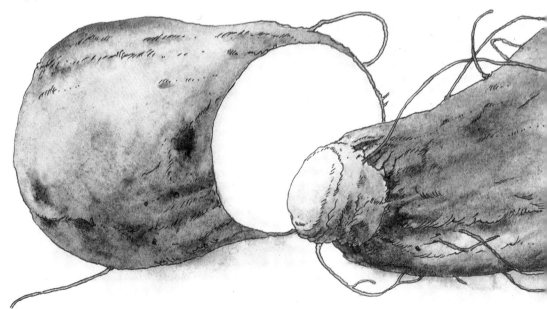

The part we eat grows underground, rather like a potato. The skins of yams are brown and rough, and rather hairy. The flesh inside is white or pinkish-brown.

Yams have a slightly sweet taste and are delicious eaten boiled, mashed, or fried.

Carrot

The green feathery tops of carrots grow above ground and look so pretty that ladies once used them to decorate their hats.

The part of the carrot we eat is the bright orange root. The root is in two parts – a hard core and a softer outside.

Carrots are sweetish to eat and can be eaten cooked or raw. People used to say that eating carrots made your hair curl and helped you to see in the dark.

Brussels Sprout

Brussels sprouts grow on thick stems about one metre tall, with floppy green leaves on top. The stems are covered in shoots. The shoots contain buds that grow into small heads or 'sprouts' – just like little cabbages. The sprouts are round and green and about four centimetres across.

Brussels sprouts are grown all over Britain, Europe and North America.

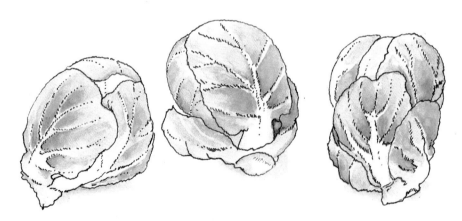

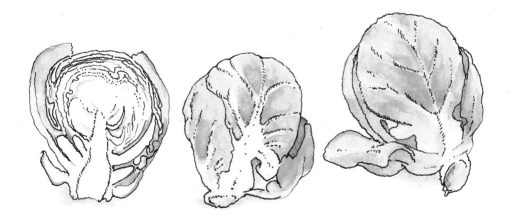

Lots of people don't like Brussels sprouts because they have a very strong taste and smell.

Aubergine

Aubergine is called 'eggplant' or 'eggfruit' in some parts of the world. This is because it is shaped like a very large egg.

It grows anywhere in the world which is warm and damp.

The aubergine bush is a low, creeping plant with pinkish flowers and a tough stem. The fruits which we eat are a shiny, purply-black colour and grow up to thirty centimetres long. They are usually round or oval, although some are long and thin.

Aubergines have a bitter taste but are delicious stewed in oil.

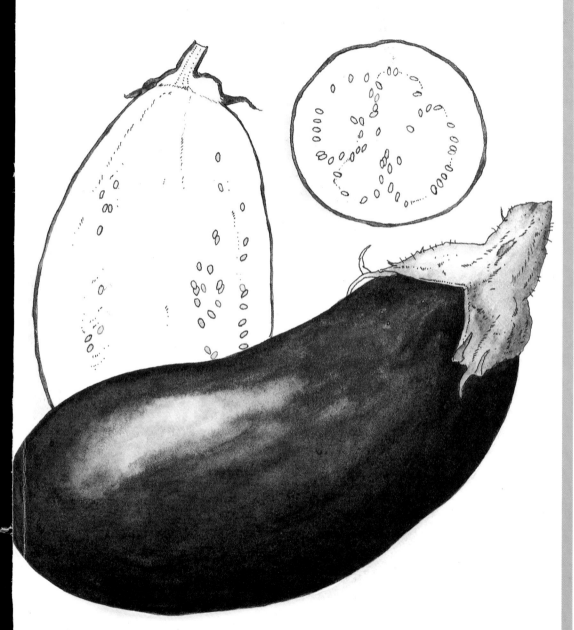

Artichoke

Artichoke plants look very like thistles. They grow up strong and tall, between one and two metres high. The leaves are grey-green and the flowers are blue.

At the top of a globe artichoke is its flower head or 'choke'.

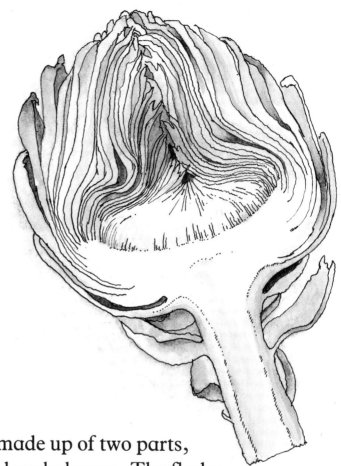

The choke is made up of two parts, a fleshy base and scaly leaves. The fleshy base is called a 'heart', and is the main part we eat – though the leaves may be eaten too, if they are tender enough.

Artichoke hearts are a great delicacy. They can be eaten hot or cold, dripping with oil or butter.

Red Pepper

Red peppers are sweet peppers – not the kind of pepper that makes you sneeze!

They first grew in the tropical parts of America, but can now grow anywhere if they are kept warm enough.

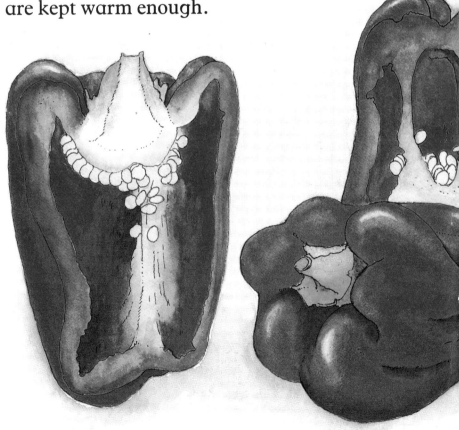

Pepper plants grow close to the ground. They have a few leaves and yellow flowers. When the fruit is ripe, it turns red and is ready to be picked.

Red peppers are sweet and juicy and can be eaten cooked, or raw in salad.

Leek

Leeks grow wild in warm countries and are grown all over Britain and Europe.

Leek plants grow upwards from shallow roots. They have a thick, round bulb which is whitish-green with long green leaves at the top. Leek flowers are round and purple.

The bulb grows mostly below ground and is the part we eat. Leeks taste rather like onions and are usually eaten cooked.

They are the national plant of Wales. There was a tradition in Wales of wearing leeks in your hat on March 1st, St David's Day!

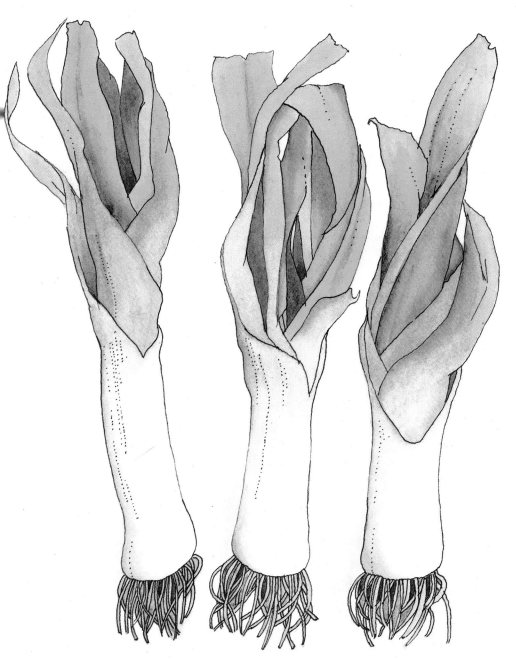

Sweetcorn

Sweetcorn is part of the maize family. It was first grown in North America, but is now grown in Europe and Britain as well. It grows on straight, tall plants. Leaves branch off the stalks, and ears, or cobs, grow in between.

The cobs are made up of lots of little yellow seeds which taste sweet when cooked. The cobs are protected from damage by long, soft hairs.

The seeds of sweetcorn are nibbled straight off the cob, dripping with butter – a messy job!

Potato

Potatoes were brought to Britain from the New World by Sir Walter Raleigh in 1586. Now potatoes are grown and eaten all over the world, and provide the main food in many countries.

Potato plants grow up to about one metre high. The flowers are white or pale-blue, and the leaves are green.

The parts of a potato we eat grow from the roots. Potatoes vary in colour from brownish-white to almost red.

They are eaten boiled, roast, fried, baked, mashed, or as chips or crisps.

Pea

Peas grow in parts of the world where it is warm and sunny.

Pea plants have hollow stems which climb up stakes, or trail along the ground. The leaves end in wiggly shoots which are used for climbing to new places. The flowers are white or purple. They are about two centimetres across.

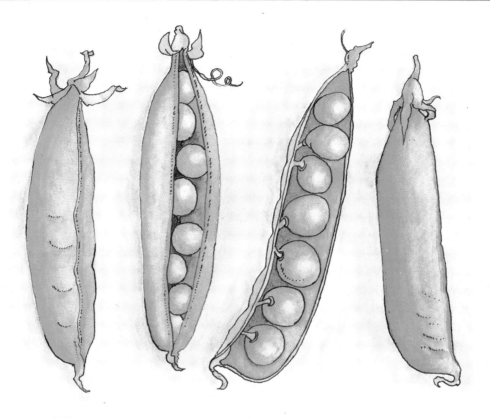

The parts we eat are the seeds, which grow in pods. The pods are green and about ten centimetres long. They split open when ripe. There are five to ten seeds in each pod.

Most peas today are picked and quickly frozen. They can be eaten fresh by taking them out of their pods and boiling them. Peas are very much alike, which gives us the saying, 'as like as two peas in a pod'.

Index

A
Africa — 7
America — 18
Artichoke — 16–17
Aubergine — 14–15

B
Britain — 12, 20, 22, 24
Brussels sprout — 12–13
Bush – aubergine — 14

C
Caribbean, The — 7
Carrot — 10–11
Central America — 8
Choke — 16, 17
Cobs — 22, 23

E
Eggfruit — 14
Eggplant — 14
Europe — 12, 20, 22

G
Globe artichoke — 16

I
India — 7

L
Lady's fingers — 6
Leek — 20–21

M
Maize — 22

N
New World, The — 24
North America — 12, 22

O
Okra   6–7
P
Pea   26–27
Pods   6, 27
Potato   24–25
R
Raleigh, Sir Walter   24
Red pepper   18–19
S
St David's Day   20
Sweetcorn   22–23
T
Tree – okra   6
Turkey   7
W
Wales   20
West Indies   8
Y
Yam   8–9